I0483758

# Reflections on Over Fifty Years in Research and Development; Some Lessons Learned

John W. Lyons

Center for Technology and National Security Policy
National Defense University

February 2012

The views expressed in this paper are those of the author and do not reflect the official policy or position of the National Defense University, Department of Defense, or U.S. Government. All information and sources for this paper were drawn from unclassified materials.

**John W. Lyons** is a Distinguished Research Fellow at the Center for Technology and National Security Policy. Dr. Lyons retired as the Director of the Army Research Laboratory and served previously as Director of the National Institute of Standards and Technology. He began his career in chemical research and development at the Monsanto Chemical Company. Dr. Lyons received his AB in chemistry from Harvard College and his AM and PhD in physical chemistry from Washington University in Saint Louis, Missouri. He is a member of the National Academy of Engineering.

**Acknowledgements:** The author is indebted to support and helpful critique of the manuscript by Drs. Richard Chait and Steve Ramberg, both of CTNSP. He also wishes to salute the hundreds of scientists and engineers with whom he worked over so many years and who contributed so much to his education as a research manager.

# CONTENTS

## LIST OF FIGURES

# REFLECTIONS ON 50 YEARS
# IN SCIENCE AND TECHNOLOGY RESEARCH

## Introduction

This paper presents some thoughts about research in science and technology (S&T) gleaned from my more than 50 years working in scientif ic and engineering research—first in th e chemical industry, then at two different government laboratories, and later some years in S&T policy. It elaborates on a paper by Richard Chait in which he interviews three fo rmer S&T executives in the Department of Defense (DOD) [1] on how to m anage a research laboratory. In this paper, I expand on the comm ents made to Dr. Chait and provide a broad context for m y discussion. In addition, this paper connects a number of subjects discussed in several other papers published by the Center for Technology and National Security Policy (C TNSP). My objective is to provide some insights on what it is like to work in a scie ntific research establishment. My hope is that this will be of some value f or the senior manager who has no laboratory experience but is responsible for overseeing a research departm ent. I also hop e the paper will help new technical personnel just entering the laboratory for the firs t time. For experienced laboratory staff, the paper will c ontain many familiar ideas and p erhaps some that are controversial. Some of the paper deals with the DOD technical programs.

I include a few anecdotes, as sidebars, of unusual and memorable experiences I had in the pursuit of my duties in the laboratory. These anecdotes are intended to show the wide variation, apart from the usual bench-top research, in the sorts of things one encounters while pursuing a career in research.

I begin in C hapter 1 with a look at the various stages of research and developm ent (R&D) and how they are connected. My focus is on the te chnical innovation process—m oving from ideas through various stages to fielded products and processes. Chapter 2 includes a description, by way of an exam ple, of the roles of a DOD laboratory, followed by a summ ary of th e characteristics of a "good laboratory." This chapter considers four kinds of research laboratories: industrial, governmental, academic, and free-sta nding institutes. My o wn experience includes industry and governm ent laboratories. Chapter 3 compares performing all work in house to collaborating with external laboratories. Chapter 4 reviews the worldwide technic al community and the various ways scientists and engineers re main informed on the lates t developments in their fields. Chapter 5 p resents my ideas on managing an R&D laboratory. Finally, Chapter 6 gives some advice to new members of the research community from my experience over the past half-century.

---

[1] Richard Chait, *Perspectives From Former Executives of the DOD Corporate Research Laboratories,* Defense and Technology Paper 59 (Washington, DC: Center for Technology and National Security Policy [CTNSP], National Defense University [NDU], March 2009).

# CHAPTER 1. R&D IN DIFFERENT ENVIRONMENTS

This chapter compares and contrasts the R&D process in different environm ents—commercial R&D, government R&D, academic research, and research in free-standing research institutes.

Science is the process of gaining understanding of the natural world. It produces theories and laws that describe the behavior of the various aspects that make up our world and our universe. Scientific research is an organized method of l earning about this behavior. A desire to push the frontiers of knowledge m otivates much of sci entific research. This research is often term ed curiosity-driven research or a search for understanding.

Engineering is the application of scientific knowledge to solve real-world problems and to design and implement creative new concepts for m anmade systems. Engineering research seeks to develop new or i mproved design and fabrication t echniques. Engineering science is a term used in some universities to describe fundam ental research into the behavior of real m anmade systems. Some fields of engineerin g research have m oved, over tim e, from engineering to science and vice versa; other fi elds have rem ained between sc ience and engine ering research. Examples of the latter fields include fluid dynam ics and the non-linear behavi or of structures in response to severe external st resses (earthquakes, high winds). Often, research in engineering uncovers gaps in scientific knowledge and thus stimulates new studies in the sciences.

A new graduate considering a career in science and engineering probably has little idea that there are different kinds of laboratory research. I didn't. The graduate will discover that both science and engineering include basic and applied research. There are laboratory opportunities in product development for new products or for improvements to existing products. Scientists and engineers are involved in research on m anufacturing processes. As product research m atures, the work shifts to advanced development—the creation of early prototypes—followed by efforts to m ove the concepts into manufacturing and technical development of the market.

After World War II, the relationships am ong these different kinds of R&D were debated, especially regarding how to move from military R&D to broader top ics in civilian science and engineering. Early rationalizations came to be known as the linear model. In the linear model, as shown in Figure 1, basic research begets applie d research begets development begets production and operations.

The linear model is a sim plification; sometimes the cart co mes before the horse. Occasion ally, basic research is performed after the applications have been developed, as happened, for

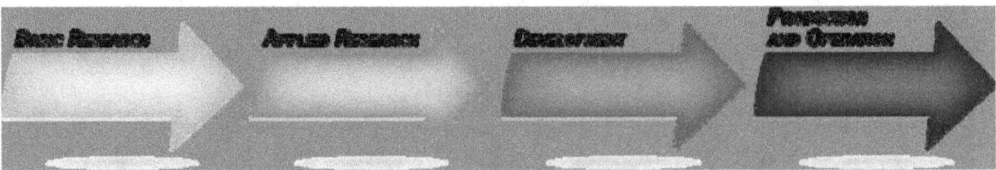

**Figure 1. Simple Linear Model of the R&D Process**

example, in therm odynamics and the steam engine. The steam engine was developed about a century before the science of therm odynamics rationalized the behavior of the engine. Metallurgy and ceram ics were practiced m illennia before we understood phase diag rams or the effects of microstructure on material properties.

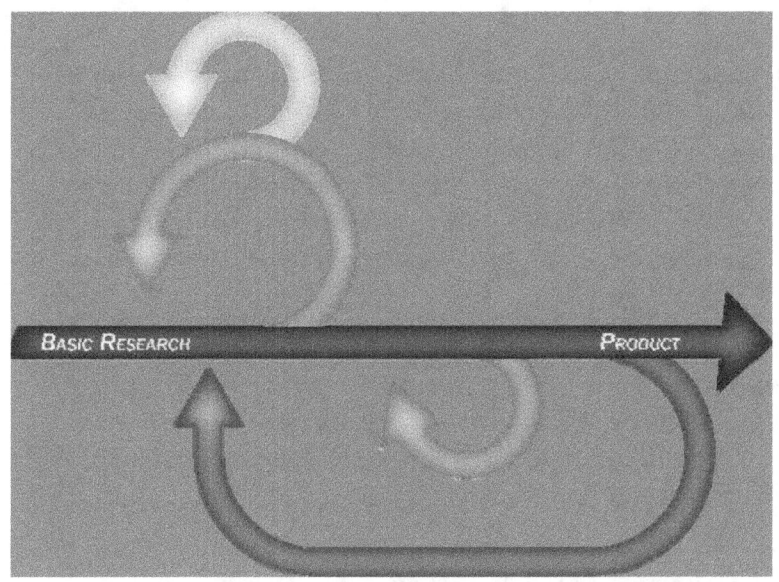

**Figure 2. Linear Timeline With Recursive Loops That Return to Earlier Phases of Research to Address Various Problems[2]**

Technical progress cannot be described in one dimension. In fact, serious knowledge gaps are often discovered during the applied research or advanced development stages. T hese gaps can only be filled by re turning to portions of the basic research stage to better understand the phenomena involved.

In very fast-m oving disciplines, such as solid state ph ysics and electronics, basic an d applied research must som etimes occur simultaneously—often in the same research group. The m odel would then show recursive loops that may go all the way from problems arising in manufacturing processes back to basic research.

Two-dimensional models are useful. The two- by-two diagram developed by D.E. Stokes is a popular model that deals with the difference between basic and applied research.[3]

The vertical axis determines the extent to which scientific curiosity with no par ticular application in mind motivates the tec hnical work. The horizontal axis indicates the extent to which solving difficult practical problems drives the work. The plot is divided into two-by-two boxes, but it could also have been drawn with the two a xes as continua such that a particular project could be located anywhere on the graph. Stokes illustrated his idea by placing noted researchers in three of the boxes. Bohr's work on the structure of the atom was curiosity driven (pure basic research). Edison's w ork was s aid to b e totally focused on the potential commercial uses of electricity (pure applied research). Pasteur's desire to treat diseases motivated him to perform funda mental or basic research studies (use-insp ired basic res earch). Much fundamental research today resembles Pasteur's—Stokes' main point.

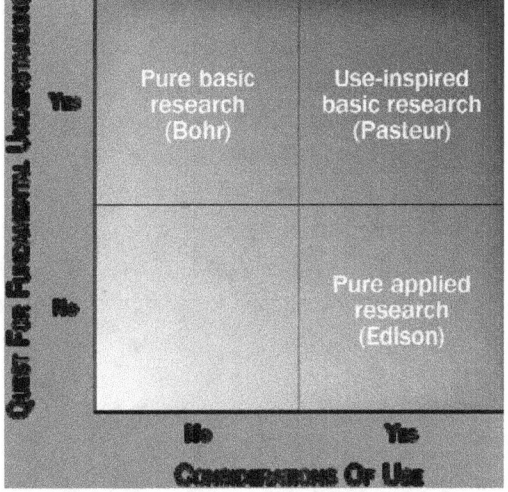

**Figure 3. Stokes' Two-Dimensional Model of Research Showing Differences in Motivation**

[2] John W. Lyons, *Two Traditions ... Whose Time Has Passed,* The Carolyn and Edward Wenk, Jr. Lecture in Technology and Public Policy, Johns Hopkins University, April 24, 1998.
[3] D.E. Stokes, *Pasteur's Quadrant—Basic Science and Technology,* Brookings Institution Press, 1997.

## Basic vs. Applied?

*In practice, use of the terms "basic research" and "applied research" is often not very meaningful. As noted in Figure 3, motivation is often the defining factor. How is one to classify a particular project ? Is it up to the individual investigator—how he or she views it—or is it up to the institution that sponsors the work (e.g., the Congress or the Pentagon)? I once heard the Chair of the National Science Board argue that the distinction is unhelpful. I agree.*

*In some Federal agencies, no official dist inction between basic and applied research exists in program planning or budgeting. For example, when I was Director of the National Institute of Standards and Technology (NIST), research was research. Each year when the National Science Foundation (NSF ) queried NIST as to how much of its work was basic resea rch, we lacked an indi cator for that metric in ou r management system. Instead, we asked senior managers for an estim ate, which we would then submit to NSF.*

*In the Army, where basic research is called 6.1 in the budget and applied research is called 6.2, I spent time in fruitless discu ssion over whether a ne w proposal was one or the other. The Pentagon's definition of basic re search holds that basic research is work that is not motivated by solving any particu lar problem. This defi nition is manife stly silly. I would wager that almost no one in the Army laboratories or the Army Research Office mounts a new program that has no utilit y for the Army in mind. Even in the case of quantum computing, we knew that su ccess would support important military requirements, such as code breaking.*

*In my own work in the Army, I tried not to use the term s "basic" a nd "applied." Rather, I found it more useful to desc ribe much basic work as "long-term, fundamental" or "exploratory" research. These terms distinguished, in my mind, whether the work addressed pressing, very di fficult problems requiring extensive and probably lengthy work or was aime d at more high-risk strategic objectives that would likely require pushing back scientific or engine ering frontiers. I felt th at exploratory research was a brief study of something new to see if we should consider a mor e serious effort. Typica lly, discretionary funds—rather than officia l programs— supported exploratory work. Long-term, f undamental work, on the other hand, was a serious, officially funded investigation of phenomena a ssociated with a formal program. The purpose of long-term , fundamental work was to remove difficult, often longstanding problems in an area—problems that required application of the most up- to-date tools and concepts.*

*I am not alone in my struggles with th ese DOD categories and not e a report from the National Research Council (NRC)[4] in which the committee r emarked: "The committee decided that discussion of that issu e [the difference between basic and applied research] is not productive, just as the distinction itself is not useful."[5]*

---

[4] *Assessment of the Department of Defense Basic Research,* Committee on DOD Basic Research, Division of Engineering and Physical Sciences, National Research Council (NRC), Washington, DC, 2005, p. 8.
[5] Ibid.

Coffey et al. recently discussed the evolution of technology with time.[6] The authors consider the development of radar by first look ing back to the e arliest scientific discoveries in theory and experiment and then m oving forward in tim e to learn what ancillary dev elopments had to occur before a radar system could be built. The stor y begins with Jam es Clerk Maxwell's theore tical treatment of electricity and magnetism published in 1865. Maxwe ll combined the two into a theory of electromagnetic radiation.[7] Following this th eory was Heinrich Hertz's demonstration in 1886 of radio waves and the fact that they can be reflected off objects. Before this information could be utilized in radar, power supplies, antennas, and similar tools were needed. It was only in the 1930s that all of these developm ents came together and a radar device could be demonstrated. It is possible to describe sim ilar chains of linked advances in nearly all technical areas.

The radar exam ple also illustrates convergence in S &T. Various pieces of the puzzle from different disciplines mature at different tim es, but they eventually conve rge to make something entirely new possible. A current case of co nvergence is that of telephony and com puting, wherein smartphones contain com puters, and com puters are a key part of communications systems. I return to the idea of convergence later in this paper after Chapter 2 describes research laboratories, what they do, and what can make them great.

---

[6] Timothy Coffey, Jill Dahlburg, and Eli Zimet, *The S&T Innovation Conundrum,* Defense & Technology Paper 17 (Washington, DC: CTNSP, NDU, August 2005).
[7] Of course, this was not really the beginning. The work of Benjamin Franklin in about 1750, Alessandro Volta in 1800, Andre-Marie Ampere in 1820, Hans Oersted in 1820, Michael Faraday in 1831, and so many others came before.

# CHAPTER 2. THE ROLES AND DESIRED CHARACTERISTICS OF RESEARCH LABORATORIES

As part of the 1991 cycle of m ilitary Base Realignment and Closure (BRAC) activities in DOD, the Congress directed that th e Secretary of Defense appoint a F ederal Commission (the "commission") to review the plans of the three military services to restructure their laboratories.[8] Final BRAC decisions on relocating or clos ing laboratories were delayed pending the commission's findings. The commission's report[9] describes the DOD labor atories' mission: "to provide the technical expertise to enable the Services to be sm art buyers and users of new and improved weapons systems and support capabilities."[10] The report lists a num ber of im portant functions, including the following:

- Acting as principal agents in maintaining the technology base
- Avoiding technological surprise
- Supporting the acquisition process
- Responding rapidly in times of urgent need or National crisis
- Supporting the user in applying emerging technology and introducing new systems.

A more recent report from the CTNSP elaborates the role of the Army laboratories[11] as follows:

- Performing in-house technical work—theory, modeling, experimentation
- Exploring new concepts and developing new knowledge
- Ferreting out, from external laboratories, new S&T of potential use to the Army
- Applying new knowledge to solve enduring Army problems
- Conducting developmental testing of new products or processes
- Conducting engineering research to aid in scale-up
- Facilitating transfer of technology to customers and users
- Providing technical advice to Army senior leadership, thereby enabling the Army to be a "smart buyer."

And that re port went o n to af firm the comm ission's list of attributes necessary for achieving high-quality, effective laboratories:

- Clear and substantive mission
- Critical mass of assigned work
- Highly competent and dedicated workforce
- Inspired, empowered, highly qualified leadership
- State-of-the-art facilities and equipment
- Effective two-way relationship with warfighters
- Strong foundation in basic research
- Management authority and flexibility

---

[8] I was appointed from my position as Director of the National Institute of Standards and Technology (NIST) to serve on the Federal Commission.

[9] *Report to the Secretary of Defense,* Federal Advisory Commission on Consolidation and Conversion of Defense Research and Development Laboratories, Office of the Secretary, DOD, Washington, DC, 1991.

[10] Ibid.

[11] John W. Lyons, Joseph N. Mait, and Dennis R. Schmidt, *Strengthening the Army R&D Program,* Defense & Technology Paper 12 (Washington, DC: CTNSP, NDU, March 2005).

- Strong linkage to universities, industry, and other government laboratories.

These roles and attributes are broadly applicable to most if not all research laboratories, not just DOD laboratories. However, not a ll of the attributes will be found in all laboratories. For example, basic research is not always a component of smaller laboratories. In recent years, some companies—and indeed entire ind ustries—have relied on universities fo r most of the basic research work of interest to the companies. Companies often sponsor such work. The following are some distinguishing characteristics of labor atories working in diffe rent parts of society—industry, government, universities, and independent research institutes.

**Industrial laboratories** operate research program s depending on the emphasis of the com pany and its customers. This work includes looking fo r new ways to increase the com pany's business and decrease costs. Customers' needs create a "demand pull" on the la boratories. These needs may be for improved existing products or fo r new products (product research). Sometimes customers simply want a lower price, thereby increasing the laboratories' focus on cost reduction in manufacturing processes (process research). Often the company is helping customers apply its products or developing new uses for its products (applications research).

### Chemical Applications in the Field

*My first task at the Monsanto Research Laboratory was to sc reen the company's products for use in treating sub-base soils under highways to overcome the tendency of clay soils to lose bearing strength when we t. After training at the then-U.S. Bureau of Public Roads in Washington, DC, the test work began. Fortunately, one of the company's products showed promise in laboratory exposure tests; two state roads laboratories were kind enough to run laboratory and field tests. Ultimately, we insta lled two te st strips in roads in Missouri and Georgia. We built our own rig for handling and applying the chemical and went out with the road crew to oversee the application; we ran soil tests as the work proceeded. (At one point on a country road in central Missouri, a local farmer appeared carrying a shotgun. He told us not to worry—that one of his large hogs*

**Figure 4. The Author Helping Build a Test Road in Central Missouri Using His Technical Findings From the Lab**

*was roaming loose. The shotgun was loaded with rock salt, which the farmer used to encourage the hog to go back to the farm. The hog obliged.)*

For those com panies that allo w their laboratories to conduc t some basic research, newly emerging technologies can open new opportunities for R&D. Alternatively, a company m ay buy its way into new areas and expect its laboratories to develop scientific and engineering support for the new area. The company makes these decisions as part of its strategic planning.

Some companies limit their overall missions very specifically; others are open to moving in to new areas. For the latter, the laboratories have latitude to investigate many new or emerging technologies. Examples certainly include the former Bell Laboratories, the IBM research laboratories, the 3M research programs, and the work of the Xerox Corporation, especially at its Palo Alto Research Center.

The authorizing legislation of **government laboratories** specifies these laboratories' mission areas—sometimes very specifically, sometimes loosely. The work actually performed depends on the appropriations process; without funds, the authority is of little value. The National Institutes of Health (NIH) laboratories are divided into 27 institutes and centers, each of which the Congress has authorized. Each has a mission to address a specified area of medical research. The Department of Energy (DOE) operates a set of "National Laboratories" that the Government owns but that operate under contract to the private sector. They each originally had a very specific mission related to nuclear weapons and energy problems. Since the end of the Cold War, these laboratories have sought additional mission areas and have broadened their scope based on their general competences in science and engineering. Compared to these laboratories, the DOD laboratories have a very focused role: DOD concentrates on research in support of the warfighter.

Some congressionally specified missions are nonetheless very flexible. For example, as one of its missions, NIST must develop the measurement methods and standard reference information needed by U.S. industry, academia, and government. As a result, when industrial emphasis shifts, NIST follows (or, one hopes, anticipates) the new needs and shifts its programs accordingly. This shift in focus is likely true of other agency laboratories, such as those in the Department of Agriculture and Food and Drug Administration.

In **university laboratories,** the focus is pretty much on the individual investigator's interests, provided funding is available. University research can be divided into two categories: (1) the work of professors and their students supported in part by grants and contracts and (2) the work of large university centers that are almost always dependent in whole or in part on sponsors from government or industry. In the first category, the professors may be funded by the university, but the students are funded by money from grants or contracts to the professors or by awards directly to individual students from government. The professor may propose a new area of work to sponsors such as NSF, DOE's Basic Energy Science Office, or NIH; if the professor wins a grant or contract, he or she is able to start on a new area. That mission area is whatever the professor wishes, depending on securing support.

**Independent research institutes** are not affiliated with any of the above categories; rather, they are set up as free-standing entities. They may be supported by very large bequests, such as the Howard Hughes Medical Institute, which is funded from its endowment. Similarly, the Salk Institute was started by support from the March of Dimes but is now also supported by the NIH and other agencies. The Battelle Memorial Institute was formed from a bequest from the owner of a steel company and initially focused on metallurgy. It moved into many other areas and greatly expanded as a result of its sponsorship of what became the Xerox Corporation. Battelle developed a contract arm and became the operator of several government-owned laboratories

(Oak Ridge, Pacific Northwest, and Brookhaven). [12] Battelle is an exce llent example of the advantages of being free from a restrictive mission statement.

**Comments.** The foregoing discussion of the various ki nds of research entities in the United States likely applies to most a dvanced countries around the world. Most of these countries have government laboratories for functions such as defense, standards, health, and energy. The laboratories may have different nam es. In Ch ina and Russia, m any leading laboratories come under their academies of science. In Japan and Europe, first-rate governm ent, university, and industrial laboratories exist much as in the United States.

The United States has bilateral and m ultilateral cooperative agreements in S&T with m any foreign countries. The DOD has b oth types of agreements. The m ost productive of these agreements facilitate ex change of in formation, exchange of scien tific staff, and occasional join t collaborations. A recent exam ple of research collaboration is the International Technology Alliance between the United States and the Unit ed Kingdom (UK) in network and information science. This alliance comprises two linked c onsortia managed by IBM—one in the UK and one in the United States—w ith members drawn fro m industry and academ ia. The ag reement ties these consortia very closely to the U.S. Ar my Research Laboratory (ARL) and the UK Ministry of Defence, which are the two sponsors of the effort. [13]

This chapter has considered the kinds of scientif ic research and the institutions that conduct or sponsor it. There are many different opportunities for technically trained people to pursue their careers. I h ave listed some of the factors one should consider when choosing a p osition in a research laboratory.

---

[12] See <http://www.battelle.org/ASSETS/36DC84C50C0049778FAE3A68E7FD1F02/75.pdf>

[13] John W. Lyons, *Army R&D Collaboration and the Role of Globalization in Research,* Defense &Technology Paper 51 (Washington, DC: CTNSP, NDU, July 2008).

# CHAPTER 3. IN-HOUSE VS. EXTRAMURAL RESEARCH; COLLABORATION

The work of a laboratory m ay be done entirely in house, or it may be contracted. If som e of the work is contracted, it m ay stand alone (separate from in-house work) or be tightly connected to an in-house effort (a fo rmal collaboration). Both approaches are used. This ch apter discusses some examples and argues for doing more collaborative research under certain conditions.

In my experience in industry, the laboratory did not sponsor m uch external work. Senior company management usually handled any exte rnal work, which involved funding universities for work not closely related to the in-house e ffort and intended to diversify the com pany. In recent years, the chemical industry has reduced or eliminated its in-house basic research and has turned to academe to perform such work.

The Government's funding of extramural research began in earnest in World War II for military purposes. After the war, the question was how to sustain the very con siderable private sector capabilities created by wartim e government funding. The first effort wa s the creation of the Office of Naval Research (ONR) established by the Congress in 1946. ONR is a funding agency and does not have a laboratory. However, in additi on to its extramural funding, it is the principal source of funds for basic and applied research at the Naval Research Laboratory. (Today, each of the military services operates a research office for extramural work. The other two ar e the Army Research Office and the Air Force Office of Scie ntific Research.) The purpose of creating ONR was to continue the program s started during th e war and to discover ne w technologies for naval warfare.

In 1945, Vannevar Bush prepared the report *Science, The Endless Frontier.*[14] This report recommended, based on the success of the m ilitary research done during World War II, that the Government should remain engaged in sponsori ng research in S&T. The report recomm ended what became NSF in 1950. NSF administers a broad program of grants to universities to promote the discovery of new knowledge and to help in the training of new inve stigators. NSF does not have a m atching in-house research laboratory. Two other m ajor grant-making agencies are DOE's Office of Science and NIH. DOE's Office of Science not only m akes grants but also manages 10 of DOE's laboratories. NIH m anages its grants separate ly from its in-house laboratories, but its laboratory staff participates in reviewing proposals.

NIST has both laboratories and extramural programs, but this ha s not always been the case. The laboratory was previously strictly in house, and staff did not have oversee external work. In the 1970s, the Congress transferred a group of university grants on fire science and engineering from NSF to NIST (which was then th e National Bureau of Standards [NBS ]) Center for Fire Research (CFR). This was not a popular m ove in the eyes of m any NBS senior m anagers, who felt that managing a portfolio of extramural programs would distract the in-house staff from their bench research. However, th e transferred programs became an in tegral part of th e in-house research; the group leaders at CFR were made responsible for overseeing the grantees' work, and

---

[14] Vannevar Bush, *Science, The Endless Frontier*, A Report to the President by Vannevar Bush, Director of the Office of Scientific Research and Development, U.S. Government Printing Office, Washington, DC, 1945. Available at <http://www.nsf.gov/od/lpa/nsf50/vbush1945 htm>.

the work was planned so as to play an important role in CFR's internal efforts. This approach created an integrated NBS-private sector center of excellence in fire research that many soon considered the best in the world. Thus, CFR demonstrated that research could be managed all of a piece with part in house and part external. This lesson has been used as a basis for subsequent initiatives at NIST and ARL.

In 1988, the Congress, in its reauthorization of NBS, changed NBS' name to NIST and added new extramural functions. One was the Advanced Technology Program (ATP), which consisted of contracts awarded to individual companies or consortia of companies large and small with or without university partners or other government laboratories. The purpose of ATP was to foster innovation by bridging the innovation gap—sometimes called the "Valley of Death"—between early laboratory findings and the more expensive advanced development necessary before going to the marketplace. A generally accepted belief was that industry and venture capital firms were not pursuing basic research results because the risks were too high. The goal of ATP was to help bear the financial burden in areas with market potential. The program was loosely patterned after the Japan Key Technology Center program created in 1985 for the same reasons.

In 1991, NIST created a new office to manage ATP. NIST laboratories did not formally collaborate with the ATP awardees but were involved in both reviewing proposals and accepting temporary assignments in the ATP office. A succession of external reviews [15] judged the program as very successful, but it became a political liability. Some in Congress believed ATP, in picking likely commercial winners, was crossing a line into the business of the private sector. Efforts were made first to restrict the program's growth and later to eliminate it. These efforts finally succeeded. In ATP's place, the Congress created a successor program called the Technology Innovation Program. [16] It is too soon to know how this program will fare.

The overall point of this discussion is that NIST has successfully managed a large extramural program while maintaining a very high-quality internal laboratory. (Three NIST staffers have won Nobel Prizes in recent years; the work of two others was prominently discussed in Nobel citations.)

ARL has found it useful to create external consortia in critical areas to fill gaps in its internal competencies. In the early 1990s, the Army decided to digitize the battlefield. ARL's challenge was to help develop and adapt the technology needed for digital communications and command and control of the battlefield. This technology needed to comprise a computerized fighting platform system of systems. ARL had not previously focused on this kind of work, but simply contracting out the necessary research would not have increased ARL's internal expertise in the long term.

In response, ARL utilized a new kind of contract called a "cooperative agreement" (not to be confused with the Cooperative Research and Development Agreements that had become popular at that time). In this approach, a consortium—led by a company with experience in building and fielding such systems—was to perform the external research. The consortium had to have at least one major research university and one historically black college or university; it could have other

---

[15] *The Advanced Technology Program: Assessing Outcomes*, Board on Science, Technology, and Economic Policy, NRC, National Academies, Washington, DC, 2001.
[16] The Technology Innovation Program at NIST was created in 2007 by the Congress in the America Competes Act.

members as well. The consortium was to be closely connected to ARL laboratories through joint program planning, and a committee chaired by a senior ARL manager was to perform overall management.

The initial funding was for three such consortia for a period of 5 years at $5 million per year per consortium.[17] To improve communication between ARL and the consortia and to speed the transfer of new findings to ARL (and to relay new Army needs to the consortia), a provision required long-term rotations of staff members to and from the consortia. The experiment began in 1995, was judged by the Army and the Congress as successful, and has been expanded over the years. The program is the Collaborative Technology Alliances (CTA). In 2001, there were five CTAs, each funded for 5 years at $7 million per year with an optional 3 years at about $3 million per year. An additional CTA was established in 2008 at $38 million for 5 years and an option of $51 million over the following 5 years. More recently, a CTA was established in network science and another in cognition and neuroergonomics. This new model for collaboration between ARL and the private sector is progressing well.

In 2006, ARL and the UK Ministry of Defence established an International Technology Alliance in network and information sciences. There are two parallel consortia—one in the UK and one in the United States. The project is funded at about $135 million for a period of up to 10 years. This project links consortia in two countries across the Atlantic Ocean. It has been called a breakthrough in international research. Whether or not such consortia will be established in other international areas is unclear.[18]

The focus in the next chapter is on the interactions among research personnel without necessarily having formal contractual arrangements. We will see that collaboration is a common characteristic of research at the level of the individual at the bench.

---

[17] E.A. Brown, *Reinventing Government Research and Development: A Status Report on Management Initiatives and Reinvention Efforts at the Army Research Laboratory,* ARL-SR-57, Army Research Laboratory, 2800 Powder Mill Road, Adelphi, MD, 20783-1197.

[18] John W. Lyons, *Army R&D Collaboration and the Role of Globalization in Research,* Defense & Technology Paper 51 (Washington, DC: CTNSP, NDU, July 2008).

# CHAPTER 4. THE INTERACTIVE ENVIRONMENT FOR THE INDIVIDUAL RESEARCHER

*"No man is an island, entire of itself..."* John Donne, *Meditation XVII*

A statement similar to Donne's can be m ade about researchers and research laboratories: No single researcher or laboratory can operate in isolation from its peers, nor could it if it wanted to. There is simply too much knowledge in the world today for any one laboratory to have command of it all, even in its own area of specialization. When considering the role and effectiveness of a research laboratory, one needs to u nderstand a laboratory's place in the broad landscape of research institutions worldwide. Except for highly classified work, res earch laboratories are connected to other laboratories by a variety of li nks. Information flies relatively freely back and forth, thereby m aking possible the early shar ing of results and enabling m any types of collaboration. These considerations suggest that practitioners in the 21st century will base their research on a m ore complete understanding of th e "state-of-the-art" and will benefit from the assistance of many colleagues around the world.

The scientist or engin eer working in a res earch laboratory receives information continuously from a wide variety of sources. By participating in this information flow, the individual becom es a contributing member of the inte rnational technical community. To be effectiv e and to avoid technical surprise, the individual must maintain external contacts, keep up with t echnical literature, attend meetings, and visit other laboratories. A way to think about this is to consider the individual researcher positioned in the cen ter of a cir cle of communication links (see Figure 5) that may also be the basis of close working partnerships.

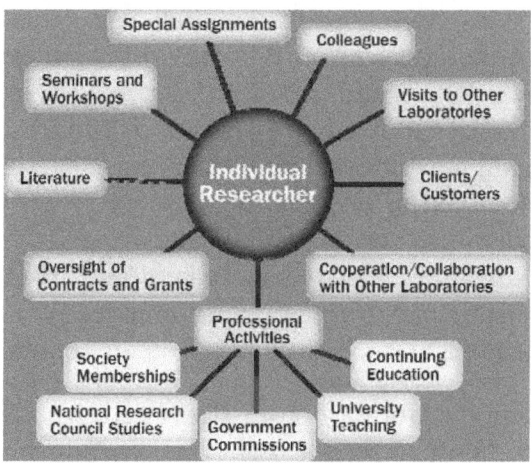

**Figure 5. Information Flows to and from an Individual Researcher**

Figure 5 shows four categ ories of contacts or activities: one-on-one personal interactions, participation in f ormal programs with o ther laboratories, technical literature review and creation, and professional activities.

Direct personal contacts include daily discu ssions with colleagues in one's own laboratory. Sometimes discussions at the lunch table lead to im portant new approache s. Sometimes whiteboards are insta lled in labo ratory corridors to facilitate inform al communication am ong staff from different lab modules. Discussions with collaborating pe rsonnel in client laboratories are a routine part of any research p rogram and are necessary to help sh ape the next steps of a program. When a formal co llaboration with other labora tories occurs, continual interactions among the personnel are essential to broaden understanding and to avoid unnecessary duplication of effort. Personnel can enhance th is interaction by traveling to the partner's laboratory or even serving a term as a visiting scien tist or engineer there. On occa sion, a scientist or engineer is assigned to oversee or m aintain cognizance of wo rk being done under contracts or grants. This assignment is an opportunity for the individual to broaden his or her ho rizons while advancing the program.

A key part of any serious resear cher's responsibility is to m aintain awareness of, and contribute to, progress in his or he r professional field. To do so, the res earcher must first keep up with the technical literature in his or her areas of interest by scanning journals and studying selecte d publications. Services that provide abstracts of publications and, in particular, online communications help this effort considerably. One can connect online to the local technical library, read the tables of conten ts of the journals, and select cer tain journals for detailed study. The actual papers may be available on line, or the journals may be in the library and available for either library study or copying. For learning about specific topics, various on line search engines are invaluable. Researchers have little excuse for not knowing what is going on in the rest of the technical world.

By the same token, the individual researcher should share his or her knowledge and help others improve their understanding by contributing his or her findings to the literature—subject to proprietary or security rest rictions. Researchers can cont ribute papers, write monographs summarizing progress in a field, give talks at technical meetings, and actively participate in seminars.

Many forms of professional activ ities contribute to inform ation sharing. Membership in professional societies brings with it subscriptions to journals, attendance at society meetings, and regular meetings of local sections of Nationa l societies. These activ ities present additional chances to listen to gues t speakers and share experiences with fellow members. Another way to gain new information is to agree to serve on special committees or commissions, which are often National or international in scope. T he NRC of the National Academ ies conducts all kinds of scientific and technological inve stigations and related policy st udies for the U.S. Governm ent. Members of study panels are drawn from a variety of sources, and interactions among committee members and government sponsors are often as valu able as the results of the study. The sam e is true for service on National studies commissioned by the U.S. Congress.

Another form of professional activ ity is developing close ties betw een local universities and the in-house research staff.. These relationships ca n include giving talks to students, serving as adjunct professors, accepting m embership on visiting boards, and enrolling in continuing education courses.

\*\*\*\*\*\*\*\*\*\*\*\*\*\*\*\*

Now, consider a group of such researchers each surrounded by all of these information channels, and then imagine connecting all of the researcher s in the world by fiber optics and the Internet, and you get som e idea of the social and tec hnical networking that occurs continuously. [19] This same network makes possible collaborative research at home and across the oceans. The Internet enables almost continual exchanges during the progress of experimental work. Collaborators can conduct theoretical studies anywhe re an Internet connection exis ts. A team of collaborators can work on a single shared software package to perform engineering design.

There are two barriers to taking advantage of th ese opportunities. One is industry reluctance to share technical information with potential com petitors; the other is the n eed to protect classified

---

[19] Thomas L. Friedman, *The World is Flat,* Farrar, Strauss, and Giroux, New York, 2006; John W. Lyons, *Army R&D Collaboration and the Role of Globalization in Research,* Defense & Technology Paper 51, (Washington, DC: CTNSP, July 2008)

information in some DOD and National security-related work. The proprietary concern is one of protecting trade secrets and intellectual property. In the case of the consortia established by ARL in the priv ate sector, each consortium is requi red to establish its o wn rules for handlin g intellectual property among m embers before biddi ng for work. The National security issue has proved harder to deal with, especially since the terrorist attacks of September 11, 20 01. Recent CTNSP publications discuss this issue.[20] The co ncern is no t with how to protect classified information—we know how to do this. The issues are in an intermediate area of unclassified but sensitive information—a gray area often labe led by the DOD as For Offi cial Use Only (FOUO). FOUO means the results cannot be published in open literature. There are costs to this control of sensitive but unclassified work. There are the actual costs of fi ghting through the bureaucracy to obtain approvals, but perhaps m ore important is the cost of lost opportunities. Nati onal Security Decision Directive 189, issued by President Reagan, declares that basic and applied research results should not b e classified except under specified conditions and that resu lts should be openly shared in the literature. Ho wever, agencies have b een reluctant to com ply with this directive.

I have discussed how laboratories can collaborate formally and, in this chapter, how individuals work together not only in their work assignm ents but also in their professional lives. In the next chapter, I present views on how to manage a laboratory or, indeed, a group of laboratories.

---

[20] John W. Lyons, *Army R&D Collaboration and the Role of Globalization in Research,* Defense & Technology Paper 51 (Washington, DC: CTNSP, NDU, July 2008); William Berry and Cheryl Loeb, *China's S&T Emergence: A Proposal for U.S. DOD-China Collaboration in Fundamental Research,* Defense & Technology Paper 47 (Washington, DC: CTNSP, NDU, March 2008). Also see *Rising Above the Gathering Storm: Energizing and Employing America for a Brighter Economic Future,* Committee on Science and Public Policy, National Academies, National Academies Press, Washington, DC, 2007, p. 105.

# CHAPTER 5. MANAGING LABORATORY RESEARCH

Chapter 5 covers lessons I have learned from managing R&D in industry and tw o government laboratories. The role of the research m anager can be thought of in term s of five different activities: planning, organizing, staffing, directing, and external relations.[21] These activities also contain additional subcategories. A summary of my experiences of these roles follows.

**Planning** begins with an agr eement as to the laboratory's fu nctions and mission. The laboratory director reviews requests and proposals for continuing, e xpanding, or shrinking existing programs and for starting new work. These proposals come from the laboratory middle managers and technical staff, as well as sug gestions from external parties. (So metimes congressional earmarks mandate work on specific topics.) The director compares the requests with the m ission and with the laboratory's capabilities and then makes a set of decisions. The decisions m ay be entirely internal, as with refocusing existing pr ograms, or m ay be external, as with presenting budget proposals for new initiatives. Over time, these decisions shape the program and develop the character of the laboratory, its effectiveness, and its reputation.

In the Government, the most rigorous planning for S&T programs occurs during the formulation of the Federal budget. Senior m anagers spend large am ounts of time preparing the budget, defending it before the several la yers of bureaucracy that must approve it, and then helping defend the budget before four con gressional committees (two authorizing comm ittees and two appropriations committees). Typically, laborato ry management works on three separate budget years at once—for mulating the next budget, defe nding the current budget before the Office of Management and Budget (OMB) and the Congre ss, and executing the pr eviously approved budget. My experience as the Director of NIST was to handle the budget presentations to the Congress, sometimes accompanied by the next leve l down of NIST m anagement. In contrast, in the DOD, my interaction with top DOD budget staff and with congressional committees was limited.

An important part of planning is determ ining the balance between short-term work and longer term, fundamental research. A part of this determ ination is deciding how much of the long-term research should go beyond the acc epted mission of the laboratory or the parent organization. This research can be the basis for expanding or , indeed, changing the nature of the parent enterprise.[22]

> *An example of considering the limits of the laboratory mission at ARL was deciding whether to fund work on quantum computing. Back in the late 1990s, this subject was at a very early stage—mostly in development of theory. One of the ARL staffers beca me interested in it and developed relationships with some external collaborators. He asked me for some of my reserve funds to continue his work. The topic was clearly outside the laboratory's current portfolio, but it represented a possible major advance in computer technology that would, per haps someday, be very important to the Army. I felt we*

---

[21] A useful general reference for this subject is Hans Mark and Arnold Levine, *The Management of Research Institutions,* NASA SP-481, NASA Scientific and Technical Information Branch, National Aeronautics and Space Administration, Washington, DC, 1984.

[22] John W. Lyons, Joseph N. Mait, and Dennis R. Schmidt, *Strengthening the Army R&D Program,* Defense & Technology Paper 12 (Washington, DC: CTNSP, NDU, March 2005).

*needed to keep a window open in this area even though we would not req uest budgeted funds to establish a formal program. So, I allocated some of my reserve funds, and the individual continued his work and maintain ed cognizance of new developments for ARL. Given the advances that have been made in this field, I think th e decision to fund the work was the right one.*

Forecasting long-term developments in S&T has not been a regular activity in the Army. The last time this was done was in 1992. [23] Recently, a more advanced techn ique has been introduced called convergence forecasting. In this appro ach, clusters of som ewhat related sciences or technologies are forecast for, say, 25 years ahead, and then the pot ential is projected for two or more disciplines com ing together (conve rging) to enable new capabilities. [24] The m anager or director of research certainly m akes use of so me form of forecasting when making decisions on the mix of program s he or she will support. Fo r me, this was an infor mal and som etimes subconscious thought process resting on a continuous stream of technical inform ation gleaned from staff, from my reading, from sitting in on meetings, and so on—activ ities that formed my basis for assessing new proposals. Som e managers rely on quantitativ e assessments based on scoring a num ber of at tributes. Some more formal technique of assessing opportunities in the future should be helpful as the pace of technological developments accelerates.

**Organizing** the laboratory to address its critical f unctions is another responsib ility of the director. Some say the shape of the organizati on should reflect the managem ent style of the director; others say the organiza tion must be structured to encourage free inform ation exchange among the staff and collaboration on project work. Both ideas are correct. My experience is that the major organizational units within the laboratory need to be bi g enough to stand on their own feet (i.e., have enough resources so that they ca n set aside reserve funds for new ideas and for survival of the occasion al budget sh ortfall). Flat organizational structures are more effective in promoting internal communications and coopera tion within the laboratory than are older hierarchical structures. The latter can create "stovepipes" or vert ical paths of responsibility and often separate functions that w ould be more effective together. For exam ple, product research is often separate from process research in industry.

A new, high-priority dem and for research that th e laboratory has not been organized to address may require the creation of new divisions or directorates. In industry, this occurred fairly often as the customer base sh ifted its inte rests. At NI ST, major new resear ch initiatives sometimes resulted in the for mation of ne w laboratory units. Exam ples are fiber optics, biotechnology, computer security, and factory autom ation. The Army's decision to digitize the battlefield led to a new directorate within ARL, as well as the creation of a new kind of external collaboration— the CTAs.[25] The Internet has enabled such new tools as the Defense Research and E ngineering Network (DREN), which was designed originally to provide high-speed connections am ong the

---

[23] *STAR 21—Strategic Technologies for the Army of the Twenty-First Century,* Board on Army Science and Technology, Commission on Engineering and Technical Systems, NRC, National Academy Press, Washington, DC, 1992; John Lyons, Richard Chait, and Jordan Willcox, *An Assessment of the Science and Technology Predictions in the Army's STAR 21 Report,* Defense & Technology Paper 50 (Washington, DC: CTNSP, NDU, July 2008).
[24] John W. Lyons, Richard Chait, and James J. Valdes, *Forecasting Science and Technology for the Department of Defense,* Defense & Technology Paper 71 (Washington, DC: CTNSP, NDU, December 2009).
[25] See the discussion in Chapter 3, page 12.

six DOD Supercomputing Resource Centers, also called Major Shared Resource Centers. Today, new groups are also emerging to deal with cybersecurity.

**Staffing.** The capabilities of the tec hnical staff represent the sing le most im portant asset and factor in laboratory success. Ensuring a top-quality staff requires an agg ressive, well-organized recruiting program with the ability to react qu ickly to hiring opportunities. A num ber of factors motivate those who enter the R&D profession—t he most important of which is usually *not* salary. Researchers do not sign on for a high salary but for the career opportunities. They want a reasonable salary and to be treat ed fairly in pay adm inistration. Strong m otivators include the nature of the m ission and the work, stim ulating colleagues, up-to-date facilities and equipment, supportive management, opportunities to collaborate with experts internal or external to the laboratory, approval to attend techni cal meetings, visits with other laboratories, and the freedom to publish or patent results.

In industry, recruiting for resear ch talent is a form al activity performed collaboratively am ong the human resources s taff and the research m anagers. Laboratories es tablish close working relationships with university placement offices and with National professional groups. Certain recruiters are authorized to m ake employment offers on the spot. In the Federal Governm ent, recruiting is much more of a challenge. In theo ry, positions are filled b y publishing notices of vacancies at the Office of Personnel Manag ement. In practice, many high-level techn ical positions are filled th rough an inform al process involving personal contacts m aintained by research staff. (Some laboratories in the Army have recently acquired direct hire authority.)

When a particularly promising individual is located, there are some other options beyond simply posting a position opening or a ttempting a direc t hire. For a can didate just completing a doctorate, one option is that the candidate app ly to the NRC's Research Associate Program ; these are usually termed "post-doc" positions. The short-term nature of such an appointm ent— usually 2 years—gives the post-doc a chance to size up the laboratory and decide whether or not to seek to stay on as a perm anent employee. Similarly, it gives the labora tory the opportunity to decide whether the post-doc is a good fit with the organiza tion. NRC only m akes post-doctoral appointments for laboratories with q ualified mentors on staff. Success as a post-doc can lead to conversion to a regu lar full-time, permanent position. (NRC's publicatio n of opportunities and subsequent screening of app licants meets the Federal requi rement for com petition.) My experience has been that up to about one-half of post-docs ultim ately convert to perm anent positions. These people have been among the very best new hires.

Sometimes one can us e regular competition for senior positions re quiring special expertise. The requirements are specific enough that only a fe w individuals can qualif y. The Government has authority, in special cases, to recruit very senior individuals with special qualifications non-competitively. However, this occu rs only rare ly. To stim ulate recruiting efforts, the resea rch director may establish performance metrics for managers to establish su ccess rates in recruiting for post-docs or for regular appointments. I used metrics in this way at ARL.

Once onboard, the new em ployee should have a m ember of the senior staff as a mentor, as well as a recently hired employee to help nego tiate local adm inistrative procedures. Retention bonuses are a way to overcom e tempting offers from outside. Laboratories should have a performance metric to evaluate managers' ability to retain the best performers.

**Managing Senior Professionals.** Any laboratory m ust have a fe w true stars who a re key to moving the work ahead and developing new areas. Some of these may be prima donnas and may require careful m anaging.[26] Some stars will d eserve special resources and labo ratory space; some should be allowed to have a staff of at least a few perm anent employees, post-doctoral fellows, and guest workers. W hen I was at Monsanto Company, a special class of "stars" was singled out by m anagement. Most had non-adm inistrative positions. They were designated "Monsanto Scientist" and "Monsanto Senior Scientist" and worked independently. At NIST, we studied the IBM Fellow s program. At the tim e, IBM Fellows were set up with 5-year funding and allowed to work on whatever interested them; a review after 5 years was required to provide the basis for continuing funding. NIST establis hed NIST Fellows and Senior NIS T Fellows. These Fellows are allotted special funding and work mostly independently. The appointments are permanent; some Fellows, on retirem ent, have been designated Emeritus Fellows. At NIST in 1997, Senior Fellow William Phillips won the Nobel Prize in ultra-low-temperature physics. He had his own group of associat es, ample funding, and excellent equipm ent. He had the enthusiastic support of his management. When asked why he stayed at NIST and resisted various recruiters, he attributed it to th e environment and to the research con tinuity he co uld achieve compared to other opportunities. (T wo other NIST physicists subsequently won Nobel prizes— Eric Cornell in 2001 for the dem onstration of Bose-Einstein condensates and John Hall in 2005 for laser-based precision spectroscopy.)

ARL has a catego ry of ARL Fellows, som e of wh om are also Arm y STs. (ST po sitions are a category of senior non-adm inistrative professionals with rank roughly e quivalent to general officers.) I required that each ARL Fellow be funde d entirely out of our appropriated base funds ("hard money") and be provided, in addition to salary, additiona l funds for their use and under their control. I thought of the ST s as rather like senior university professors, often with a group of graduate students or post-docs. I feel the STs should be granted considerable independence and opportunities to ex plore new areas. The A rmy only is authorized about 50 S T positions. There ought to be more.

Given that these appointments are based on many years of high-level performance, the incidence of poor performance has been, to my knowledge, nil.

**Directing** the laboratory means ensuring the people, e quipment, facilities, and support services are in place. It means ensuring the work is in acco rd with research plan s or that said plans are adjusted as research fin dings may justify. It means encouraging and motivating the staff. The director should be flexible in how he or sh e deals with personnel. So me personnel will need continual encouragement; others may only occas ionally need to be energized. All requ ire personal attention. W alking around the laboratory and hearing informally about new results is one of the great pleasures of serving as research director. I tried to se t aside special times for these tours, pushing aside the distractions that are always present.

The research director is responsible for all aspects of the laboratory. The director must learn how to assimilate a great deal of inform ation and di still it into manageable in formation packets that

---

[26] An anecdote from baseball: When Joe McCarthy moved from manager of the Yankees to the Red Sox in the 1940s, he was asked how he would get along with Ted Williams. Ted had something of a reputation as being difficult, especially with the press and some fans. Joe's reply was to the effect that if he couldn't get along with Ted Williams, then he shouldn't be a manager in major league baseball.

are useful as issues aris e. This ability to abso rb the essence of a pres entation is particula rly important when the subject is technical. Clearly, the director cannot know as much as the individual staff experts, but he or she m ust learn to grasp the core argument in just a f ew minutes. Occasionally the argument is unclear, and an additional briefing may be in order. If the subject is unclear, the fault may lie with the br iefer. Some specialists are particularly gifted at compressing a com plex subject into read ily understood summaries; others are not. I had to negotiate around these characteri stics and often had to study independently on m y own tim e. Briefings are m ost productive when given in th e laboratory m odules where the work is being done.

The director should delegate authority and responsi bility as much as possi ble. Support functions are critical to the success of th e technical work; a very capable head of adm inistration should be in charge. The director's staff handling budget preparation and monitoring program progress are important. Most of the director's immediate st aff should have had prior experience in the laboratory to counter the common complaint that the staff around th e director do not understand the challenges staff in the laboratories face.

The director should have a discretionary fund for supporting im aginative new ideas from the staff. This fund should be drawn from the ba se funding at levels from 2–3 percent up to 10 percent. A report of the W hite House Science Council's Federal Labora tory Review in 1983 suggests the latter figure.[27] (Known as the first Packard rep ort, this report was the result of a series of visits by a panel appointed by the Council's chair to look into issues at the laboratories.) I created a director's fund, annua lly issued calls for proposals from the staff, reviewed the proposals with senior technica l managers, and funded a f ew of the proposals for an initial exploratory phase—usually with enough m oney to support one professional and perhaps an assistant for 1–2 years. If the work dem onstrated promise, I could then establish a new progra m and incorporate it into the budget. The results included extension of existing program scope and some entries into new areas.

Directing also includ es overseeing management functions, includi ng the fiscal side, to ensure compliance and solvency. Som e would call this a separate function, or contro lling. Theories of management sometimes separate the management functions from the leadership functions. I find that effective laborato ry directors are both good leaders and good m anagers. The leadership function is more hands-on in term s of m aking technical decisions; the m anagement function is more of overseeing how administrative staff handle day-to-day operations.

**Assessing Laboratory Quality—Peer Review.** To ensure high quality in a research laboratory, some form of independent, unbiased peer review is necessary. A recen t report[28] discusses how such a review should be done. For tim eliness and re levance, the user of the research results is best suited to evaluate the work—either the work solves the user's problems in a timely way or it does not. F or technical quality, three som ewhat different types of assessm ent are possible. External peers who are experts in their fields can evaluate scientific proposals. Grant-making entities, such as NSF, NIH, and the three D OD research offices, routinely perfor m external

---

[27] *Report of the White House Science Council Federal Laboratory Review Panel,* Office of Science and Technology Policy, The White House, Washington, DC, May 1983.
[28] John W. Lyons and Richard Chait, *Strengthening Technical Peer Review at the Army S&T Laboratories,* Defense & Technology Paper 58 (Washington, DC: CTNSP, NDU, March 2009).

proposal review. Peer-reviewed jo urnals can eval uate finished pieces of work and, in tu  rn, submit the manuscripts to acknowledged experts.

To evaluate a laboratory's current work as a whole, it is necessary to assem ble a panel(s) of experts from outside the labor atory. The panels visit the laboratory, listen to project presentations, tour the laborator y facilities, and sit down w ith management and research staff members. It is b est, in the evaluatio n contract, to specify who will m anage the review, how to address conflicts of interest, and the role of the laboratory managers. A body such as NRC has the advantage of being able to convene top ex perts and command respect for the c redibility of the evaluation report. N RC insists on controlling the details a nd the conduct of the evaluation and has explicit rules for handling conflicts of interest.

By using external peer review, the laboratory not only obtains ex pert, critical assessm ents but also broadens its technical understanding through staff interactions with the panel m embers assembled by NRC or another contractor

### *Peer Review of NIST and ARL*

*I first became involved w ith NBS when asked to serve on an ad hoc NRC panel to evaluate the Bureau's various programs in research on fire. NRC has operated peer assessments of the NBS/NIST laboratori es for about 50 years. This panel on fire research made a series of recommendations th at resulted in a substantial reorientation and consolidation of the work. The experienc e resulted in my leaving industry and taking a po sition to head up the N BS fire p rogram. The NRC panel on fire research became a standing panel, meeting for assessments every year.*

*I established a similar set of NRC panels to assess the work of ARL under the NRC ARL Technical Assessment Board. This Board continues to conduct peer reviews of the ARL program. The reviews have been very helpful in pointing out potential areas of concern and generally strengthening the laboratory's work.*[29]

**Performing External Relations.** The head of the laboratory has many responsibilities: to the management chain above the lab, to the m any clients and users of the laboratory's results, to the Congress, and to the general public. The director is, to a considerable extent, the external face of the laboratory.

As Director of NIST, I handled interactions with the Secretary of Commerce and the Commerce Undersecretary for Technology, with OMB, a nd with congressional comm ittees. The congressional appearances dealt primarily with authorization and budget issues, but som etimes they were about specific investigations. For example, NIST had som e responsibilities in building construction codes and building saf ety. Some of this involved investigating building collapses during construction and building failures during earthquakes. Notable tes timonies were on the earthquakes in Mexico City in 1985 and San Francisco (Loma Prieta) in 1989. Another was a court-mandated study of a device reputed to produ ce more energy than was put in. (It didn't; see the sidebar below.) I testified on the perform ance of a copy protection schem e for digital tape recordings. (NBS questioned its re liability.) NBS m anagement called these inv estigations "hot

---

[29] NRC recently set up the Laboratory Assessment Board (LAB) and placed the NIST and ARL assessments under it. Additional laboratories have begun such assessments under LAB. The author is currently the chair of LAB.

potatoes" because of the great public interest an d the liability issues often involved. NBS was very careful as to what could be concluded a nd how the findings were ex pressed. A recent NIST example is the m ajor investigation of the Septem ber 11, 2001, collapses at the World Trade Center.[30]

### Defending the Laws of Thermodynamics

*NBS/NIST has long been used as a court of last resort in making difficult or controversial measurements. During my tenure, a number of very challenging assignments came to the laboratory, us ually from the D epartment of Commerce, members of Congress, or other Federal agencies. On at least two occasions, the work was done to comply w ith proceedings in F ederal court. O ne such example was the*

*evaluation of an electri cal device purported to produce more energy output than the energy input. The U.S. Patent Office had rejected an application for the device based on the laws of physics; namely, the law of conservation of energy and the second law of thermodynamics, which holds that perpetual motion devices are not possible. On appeal to Federal court, the Patent Office was directed to have NBS evaluate the cla ims of the inventor. NBS conducted, and reported in 1986, very careful measurements with special equipment not found in most laboratories. The laboratory in which the work was done was kept under lock and key, and the materials and equipmen t treated so as to maintain a clear trail of evidence. The upshot was that the device did not perform as claimed: "At all conditions tested, the input power exceeded the output power. That is, the device did not deliver more energy than it used."[32]*

*NBS provided the results to the Patent Offic e and thence to the Federal court. I testified to a committee of the U.S. Senate. Nonetheless the inventor continued for many years to promote his device, without success.*

**Figure 6. Joseph Newman's Energy Device. On the left is a rotating magnet; on the right, a very large copper coil. The setup connects to a large battery pack and a resistive load.[31]**

---

[30] For a lengthy series of references to NIST reports on the collapse of the World Trade Center on September 11, 2001, go to <http://wtc nist.gov/NCSTAR1>.

[31] Ibid.

[32] Robert E. Hebner, Gerard N. Stenbakken, and David L. Hillhouse, *Report of Tests on Joseph Newman's Device,* National Bureau of Standards Information Report 86-3405, National Bureau of Standards, Gaithersburg, MD, June 1986.

At ARL, relations with warfighter representatives were very important. ARL conducted visits and presentations at various sites of the Army Training and Doctrine Command to discuss soldiers' needs and recent research results. I reported to the four-star commanding general of the Army Materiel Command (AMC). In this role, I attended AMC command meetings at various AMC locations and participated in command group deliberations. I made presentations to many of AMC's suppliers as well. In budget and planning activities, I worked closely with the Deputy Assistant Secretary of the Army for Research and Technology and sometimes with his supervisor, the Assistant Secretary, who also was the Army Acquisition Executive. On occasion, I presented ARL work to the Director, Defense Research and Engineering

## In the Field With ARL

*The work of researchers is usually fairly quiet and involves having a tolerance for long hours of careful and often repetitive experimentation, much of which fails for one reason or another and must be done over. The work of the research manager is equally fairly routine. However, on occasion something unusual occurs that leavens the day's work. Here is an example from my experiences as director of ARL.*

*My deputy director was a full colonel in the armor branch. He felt I couldn't be a proper manager of Army technology without intimate knowledge of Army combat platforms—especially the Abrams main battle tank. Aberdeen Proving Ground has extensive facilities for testing tanks, including test tracks for determining maneuverability and structural stability and also firing ranges for evaluating the tank's armaments. The colonel arranged for me to drive an Abrams on the track and fire the tank's main gun on the range. On the appointed day, I was dressed appropriately to drive the Abrams. The outfit included a helmet fitted with earphones. The driver's compartment is isolated from that of the tank commander, and communication between the two must be by radio. I observed this with some apprehension recalling that Michael Dukakis, candidate for president in 1988, was similarly photographed rising from the driver's seat of an Abrams with the helmet on. This picture somehow made Dukakis an object of derision by the Republican opposition. I cautioned the colonel and our entourage not to photograph me in a similar position. They did take such a picture and promptly published it in the ARL newspaper. I still have the framed photo as a treasured memento. Incidentally, the Abrams, with its 1,500 horsepower turbine engine, is smoother and easier to drive than most smaller vehicles.*

*The second part of this initiation was to go to a firing range at Aberdeen to fire the Abrams 120 mm main gun. This was the gun that dominated Desert Storm when supplied with the kinetic energy round known as the "silver bullet." So I climbed into the gunner's seat, was familiarized with the gun sight and trigger mechanism, and indicated I was ready. Through the sight, I focused on a target way down range, but I found that the target kept floating up out of view. I had to track it and guess when to fire. Needless to say, I didn't hit the target with any of several shots. I later learned that the gun's system for compensating for the effects of gravity was not turned on. I did gain a healthy respect for the gun in terms of noise and recoil. A considerable headache ensued.*

**Figure 7. The Author in the Driver's Compartment of an Abrams Main Battle Tank on the Test Track at Aberdeen Proving Ground**

All in all, the research director has many responsibilities outside the confines of the laboratory facility; the laboratory's staff help carry these responsibilities out.

# CHAPTER 6. RECOMMENDATIONS ON LABORATORY MANAGEMENT

Reflecting on my experiences at three different re search institutions reveals a number of lessons widely applicable to any laboratory.

The laboratory at the chem ical company was de voted to developing new products and new us es for existing products f or a variety of industrial customers. Associated with this activity was process development for m anufacturing new pr oducts and cost reduction of processes for existing products.

At NIST, the client base consists of scien tists and engineers around the world who rely on standards for measurements, standard reference data, and standard reference materials. The entire technical community relies on NIST for these servi ces. NIST is the Nation's cou rt of last re sort for physical and chemical measurements and, increasingly, for biological measurements. NIST's sister laboratories in other countries are not so much com petitors but colleagues in obtaining and disseminating the best measurements.[33]

ARL is more like an industrial laboratory than a typical governm ent laboratory; it is much more sharply focused. ARL is the central laboratory for Army materiel. Its responsibility is to conduct basic and applied research in support of the Arm y acquisition comm unity's efforts. The wor k product of ARL is transferred to a set of engineering centers that carry out advanced development and then provide technical support to Ar my acquisition product m anagers. These product managers, in turn, com plete the innovation cycle by converting prototype new product s into fielded Army materiel.

Despite these differences, the lessons learned from managing these three laboratories are similar and applicable across the laboratories.

### 1. It is essential to establish general agreement on the laboratory's mission.

One must ensure the mission is consistent with that of the parent organization. Also, the manager should determ ine how m uch flexibility exists for expanding the m ission into new areas. As noted in Chapter 1, some organizations are very flexible, and some are quite firm in holding to the present mission.

### 2. Continuous long-range planning is important.

All laboratories have a planni ng function. A long-range plan is important to guide the preparations for future work. It should be the basis for the annual budget subm ission, which also serves as a short-range plan.

In the Army, the Office of the Chief Scientist oversees the preparation of the Army Science and Technology Master Plan. The focus is on implementing research program s already agreed upon rather than c onsidering long-range opportun ities. Successful long-range planning requires forecasting developm ents in research over, say, 25 years, and forecasting

---

[33] These National laboratories address the responsibilities for measurements that fall under the Treaty of the Meter signed in 1875. Their work is overseen by the International Committee on Weights and Measures and its Bureau of Weights and Measures, with offices at the Bureau International des Poids et Mesures' laboratory in Paris, France.

likely capabilities needed by the warfighter over the sam e period. The result will represent the vision for the S&T program.

### 3. *The organization of the laboratory should be consistent with the management style of the director and should facilitate communication and cooperation within the laboratory.*

Many organizational changes are m inor, affecting only sm all segments of the laboratory. Major reorganizations usually alter a laboratory's overall structure and affect everyone. Such full reorganizations should not occur t oo often. NBS/NIST saw only four m ajor reorganizations in the first 90 years of its existence.

As noted in Chapter 3, laborat ories are now tending to have flatter organizations with le ss hierarchy. The hope is that this structure will be m ore efficient in term s of internal communications.

### 4. *A highly competent staff is the most important factor in the success of a research laboratory.*

A laboratory must use all avenue s of attracting and retaining excellent staff. Some options, such as post-doctoral appointm ents, are very effective in adding new staff. Mentoring is also important. Managers should make every effort to retain top employees.

### 5. *The director must be involved with all the staff and insist that subordinates stay close to the bench-level personnel.*

The director needs to take adv antage of all expertise in the laboratory, including the director's senior associates and subject m atter experts ac ross the lab oratory. Particularly useful advice is also available from non-managerial senior staff, known in the Arm y as STs or Senior/Technical Professionals.

The laboratory should have a discretionary di rector's fund to suppor t new ideas proposed primarily from the staff. This fund c an be as little as 3 percent of base funding up to as m uch as 10 percent (see Chapter 5).

### 6. *Outside experts should perform regular, formal, and external reviews of a laboratory's quality.*

An external group of experts should assess a laboratory's quality regularly. An external body, such as a contractor, should m anage and conduct the assessment. Acknowledged, independent experts should perf orm this unbiased appraisal. Ca re should be taken to handle conflicts of interest. This review is hard to do in industry b ecause so much of a lab oratory's work is proprietary. Often, external consultant s review specific portions of a laboratory's work.

# APPENDIX A. THE RESEARCH LABORATORY—THEN AND NOW; SUGGESTIONS FOR THOSE ENTERING UPON AN R&D CAREER

When I was an undergraduate m ajor in chem istry in the late 1940s, we di dn't do research or write an undergraduate thesis. Experim entation was closely specified in the various laboratory segments of the coursework. All of the bench work was hands-on synthesis and analysis usin g only the simplest of instrumentation, such as weighing and temperature measurements.

When I arrived in the research department in a division of the Monsanto Chemical Company in 1955, research experim ents made use of X-ra y crystallography, som e spectroscopy, earl y versions of chrom atography, high-temperature o vens and furnaces, auto matic titration devices, and pH meters. We were initially at a remote site and, to obtain X-ray information, had to send samples halfway across the country to a com pany laboratory with that capa bility. Most of what we did was weighing, h eating, dissolving, prec ipitating, extracting, and m aking titrations of various kinds. When I moved to a new laboratory facility at a new company headquarters, the instrumentation was consolidated. The laborat ory had ap plied recently developed nuclear magnetic resonance (nmr) techniques using early, commercial nmr spectrometers to elucidate the structure of complex phosphorus com pounds. This analytical tool becam e a standard experimental technique to com plement the va rious X-ray and spectroscopic devices at our disposal. However, these machines were set up in separate laboratory modules and monitored by specialists. These folks trained users, kept the machines in operation, and did their own research. (Note: These practices were for chem ical research laboratories. The situation would have been different for different disciplines.)

Research in those days was cons idered to be either in th eory or in experiment; sometimes, research was sim ply observation, as in astrono my or biological taxonomy. The advent of the computer changed all of that. In the m id-1950s, Monsanto acquired an IBM 701 computer and made it available to the staff. Howe ver, the computer group generally did the programming, and the researcher m erely provided the problem and some data. A select group of specialists— mathematicians, statisticians, and some engineers—programmed and operated the computer. The individual researcher designed and ran experim ents and t hen took the data to the com puter group. Later, the computer group sent a printout of results back to the researcher. Monsanto had also acquired a large an alog computer useful in m odeling process operations. I persuaded the chemical engineers at the analog machine to model a wet-process cem ent kiln (steady state) for use in a technical support program on which I was engaged. Once the kiln was modeled, we tried various input conditions to see if the postulated effect on performance could be realized. It could. They went on to produce a digital model of the tim e-dependent behavior of a kiln. This was m y first encounter with modeling and simulation. The researcher generally did no programm ing and had no direct interface with the computer. Not until years later was remote access provided from terminals distributed throughout the com pany's laboratories. Then such use m eant researchers either had to write their own program s (usually in Fortran) or had acquired program s from someone else. By the tim e I left in 1973, there were still no com puters in the research lab modules.

This all changed rapidly. W hen I went to NB S in 1973, wave after wave of new de velopments were occurring in so lid state dev ices and com puters. First came the s o-called minicomputers, such as the DEC VAXs, that were set up in        a few laboratories at NBS. Small groups of researchers used, operated, and maintained these minicomputers. Additional staff were not hired just to maintain and operate the computers. A number of these machines were installed for special applications, such as designing circuits for silicon chips and devices or modeling chemical reactions at the molecular level. Soon, the personal workstation and the personal computer (PC) were introduced.

At about this time, the staff in the computer institute developed NBSNET to provide access by wire around the campus. This was one of the first local area networks. NBS secretaries turned in their IBM Selectric typewriters for the early PCs. NBS standardized use of these computers so report text and other documents could be moved around the lab from one machine to another. In a parallel development, small computers began to be installed in experimental apparatus to monitor and control experiments. Vast amounts of data were easily collected and subsequently analyzed. Lab benches began to look like those in electronics or computer companies.

Ultimately, nearly all scientific equipment became digitized, providing controlled introduction of reactants, controlled changes in conditions, and automated measurement of results. The results were often analyzed during the experiment, and answers were printed out. Thus, one could analyze something with a mass spectrometer, and the machine would compare the output to tables of data stored in its computer and then print out the most likely compound. The laboratories became automated. Researchers had to learn a great deal about these machines and their nuances and less about manipulating experimental materials.

When the power of main computers reached a certain level, a new third branch of research was recognized—namely, computer modeling and simulation. With the supercomputer, a physical and chemical model could be built for various problems. Once the model's performance was verified by comparison with actual experiments, the computer model could be used to make many runs to test the sensitivity of the phenomenon being studied to changes in parameters. For some problems where running many full-scale experiments would be too costly, the computer models enabled much broader and more informative studies. Today, many such models can be developed and run on powerful desktop computers. But for the most complex modeling, the largest and fastest supercomputers are still, and probably always will be, needed.[34]

In the 1940s, 1950s, and even into the 1960s, the research chemist's bench top consisted mostly of reagents on shelves or inside a fume hood, lots of lab glassware, a few stirrers and hotplates, and an occasional pH meter or automatic titrimeter. One of the regular tasks, for professionals as well as lab assistants, was to carefully wash and rinse the glassware, sometimes cleaning it first with baths of hot, concentrated acids. Residues of experiments were almost always flushed down the sink, relying on a neutralizing basin outside the building. No more. Today, laboratory waste is classified as hazardous, placed in containers, and hauled away by special contractors to special disposal sites.

Today, when walking through a chemistry laboratory, one may not see many lab modules with reagents on shelves and glassware on the bench. Now, a chemical lab or a physical chemistry lab

---

[34] A recent NRC study made this point. See *The Potential Impact of High-End Capability Computing on Four Illustrative Fields of Science and Engineering,* Division on Engineering and Physical Sciences and Division on Earth and Life Sciences, NRC of the National Academies, National Academies Press, Washington, DC, 2008.

often consists of bench-top devices that m ay each cost $1 million or m ore. These m ay be analytical instruments, devices for s ynthesizing systems such as m olecular beam epitaxy for microchip research, and system s for observing be havior. These devices m ay operate at extrem e pressure (or vacuum) and tem perature (very low or very high). I have seen individual research lab modules filled to b ursting with machines operating at very h igh vacuums and ultralow temperatures—not just one but se veral in a m odule. In an NBS laboratory that studies the behavior of trapped ions at near absolute zero temperatures, it was difficult to move around because the space was so full of apparatus. Some labs use high vacuu m systems to study th e behavior of a few atom s or molecules moving about on substrate surfaces. Lasers are ubiquitous and used for measuring lengths, defining time and frequency, burning patterns into surfaces, and on and on. An interesting phenom enon is that if an experiment originally requiring a roomful of apparatus were to b ecome of comm ercial interest, a company would eventually find a way to miniaturize the system , even to m ake it po rtable. This happened with m ass spectrometers and other tools used in the field to make measurements.

Some experimental systems require special facilities. A research reactor, for exam ple, requires a confinement chamber for the reactor itself and an associated instrument hall where b eams from the reactor are incorporated into particular experimental setups. The reactor a t the NIST Center for Neutron Research, for instance, has nine instrument stations in the confinement chamber and an additional fourteen stations in an adjacent hall where b eams of neutrons are guided to experimental stations. Each station performs a particular experimental task. A continuous stream of guests from NIST program s and other privat e and public research or ganizations use these stations for a few days at a time. A synchrotron for providing beams of well-characterized far ultraviolet and X-ray radiation re quires special support services. Se veral "light sources" of this nature exist around the world as designated user facilities. Everyday scientific and engineering research has become an expensive and often very complicated endeavor—far more than 50 years ago.

\* \* \* \* \* \* \* \* \* \* \* \* \* \* \* \*

In thinking back, I can determ ine some factors that are important for a fulfilling career in R&D. As a young person just thinking about such a career, good preparat ion requires a few considerations. I decided to become a chemist and majored in chemistry in college. In addition to courses in chem istry, I took 2 years of calculus and 1 year of physics. That was clearly not enough.

I finished work on my bachelor's degree, put in 2 years in the U.S. Army, and then went to work in the research departm ent of a division of th e Monsanto Chemical Company. I quickly learned that an AB in chem istry wasn't enough for a car eer in research. So I took graduate courses and did my thesis, leading to a PhD in physical chemistry. My PhD provided a background in thermodynamics, statistical m echanics, polymer science, quantum chemistry (which didn't help much when quantum computing came along), and ot her courses I felt I needed. But I still didn't take enough physics and math.

I found, after tak ing positions in research m anagement at NBS (later NIST) and later at ARL, that I was overseeing work in physics, material science, and engineering, as well as managing the computing enterprise. For example, I needed to understand the engineer's approach, particularly for open systems. Chemists largely deal with closed systems in equilibrium. Engineers deal with open systems involving heat and m ass transfer. These systems require a different way of addressing problems and a different m athematical approach. I had no for mal training in these areas and had to study the subject m atter on my own. The same was true in physics. At NIST, I had to oversee som e very fundam ental research in physics. I learned most of it through m any briefings from the scientists. Som e of this I filled in by independe ntly studying textbooks on physics, such as Richard Feynm an's Lectures in Physics and the Berkeley Course in Physics. These were helpful, but I would have benefi ted more from classroom study. I strongly urge people entering scientific work to take as m uch math and physics as possible—they underlie nearly all of the physical and biological sciences.

In addition, before starting serious study for a research career, som e summer work in a laboratory would be helpful in und erstanding what the wo rk entails and in deciding whether to go to graduate school. I believe a career in research requires a Ph D in a relevant discipline. I didn't go to graduate school until I was working full-time in a laboratory and had already started my family. The burden of two jobs, a fa mily, classwork, and thesis research was ex hausting; I would not advise anyone do it this way.

Once at work in the laboratory, I recommend that one seek out a m entor or sponsor—someone you respect and who can get things done in th e organization. This m ore senior person m ay providing coaching in specific challenges to you or may offer more general philosophical advice. I found such a person at Monsanto. I never worked for him, but I learned from his example and received the career guidance I needed. W hen we were talking about the future, he suggested the following exercise: W rite down the 10 m ost important things you want to accom plish, and then write down the 10 things you are willing to give up to achieve those accomplishments. I tried it and couldn't get beyond about five item s in each category. This analytical approach led to m y return to graduate school and la ter to my leaving the company and joining the U.S. G overnment at NBS. I've never regretted it.

Another thing I learned from this individual is to "always leav e tracks." By this, he m eant writing and speaking before audiences and being active in prof essional activities outside one's immediate environment. He was a prodigious contributor to scientific literature, including a great many scientific papers and books and two treatis es on the chemistry and uses of phos phorus and its compounds. He was known as perhaps the world's expert in his field. He went on to serve as a senior professor of chem istry at a lead ing university. By leaving tr acks, one estab lishes a reputation both within one's ow n organization and within the community a t large. When it comes time to make a change, one's external reputation is a critical factor in employability.

When deciding how to publish findings—interna l report or outside ref ereed publication—the people I respected m ost usually opted for external publication, assum ing the company would clear it. I published both internal reports and exte rnal refereed articles, and I authored or co-authored three books. My m ove to NBS ca me as a result of a m onograph (book) I wrote on the chemistry and uses of fire retardants. It was one of the first treatises on this subject and received considerable attention. Just afte r its publication, NBS was looking in to its work on fire safety standards as a result of congressional interest. NBS requested a study by NRC, and NRC created an ad hoc comm ittee for th is purpose. The co mmittee had a num ber of academ ic experts in physical modeling of combustion phenom ena and n eeded someone to re present chemistry. My book caused NRC to put me on the committee. My service led to NBS recruiting m e to come to

NBS to manage a new, consolidated NBS Center for Fire Research. I decided to do it for little while; I stayed for 20 years.

Another lesson I learned early on was the value of taking a fundamental scientific approach to what would appear to some as a mere technical service. The best scientists and engineers are those who look deeply into a problem —whether it arises in new research, the production department, or the sales department. These scientists and engineers apply their knowledge of basic science and engineering to find a solution. Often because of this approach, they can contribute to the community at large by publishing non-proprietary results. This is a lesson I have observed in practice throughout my many years in R&D. The finest chemical process engineer I ever met always began a problem by developing the thermodynamics, kinetics, and equations for heat and mass flows for the operations in question. When my colleagues agreed to model the wet-process cement kilns I worked with, they performed what we now call physics-based modeling. The modelers required thermodynamic and kinetics information, along with the characteristics of the kiln and the combustor, to enter into the analog computer—and later the digital computer.

I later expanded this lesson on the importance of taking a fundamental approach to research work to the work of a group, division, or entire laboratory. I believe an effective laboratory must have a strong component of basic research on which to build mission-oriented work. Indeed, I think about 15 percent of a laboratory's work should be in this category. It turns out, many others agree with this view.

www.ingramcontent.com/pod-product-compliance
Lightning Source LLC
Chambersburg PA
CBHW081412170526

45166CB00010B/3306